HOW
TO
NAVIGATE
OUR
UNIVERSE

MARY SOON LEE

How to Navigate Our Universe

Copyright © 2023 by Mary Soon Lee

All rights reserved. No part of this publication may be reproduced in any form, or by any means, without prior permission in writing from the author, except for the use of brief quotations in book reviews or critical articles.

ISBN 979-8-9885922-1-1

Poems from this collection first appeared in the following publications: Abyss & Apex, Amazing Stories, Analog Science Fiction & Fact, Andromeda Spaceways, Apex Magazine, Asimov's Science Fiction, Dreams & Nightmares, Eternal Haunted Summer, Eye to the Telescope, Frozen Wavelets, Louisiana Literature, The Magazine of Fantasy & Science Fiction, Mithila Review, NewMyths.com, Paper Street Online, Penumbric, Rune, Silver Blade, Songs of Eretz Poetry Review, Spillway, Star*Line, Strange Horizons, Triangulation: Dark Skies, Twitter, Uppagus.

Cover art copyright © 2023 by Lucy Lee-Moore

Cover & interior design by Jay O'Connell

To Lucy and William, I will always love you. Always. Always. Even if I'm on the other side of an event horizon.

And to Pluto, you will always be a planet to me.

PART I OUR BACKYARD: THE SOLAR SYSTEM

How to Paint Mercury........................... 1
How to View Venus.............................. 2
How to Typecast Venus.......................... 3
How to Colonize Venus 4
How to Thank Earth 5
How to Decorate the Moon 6
How to Witness the Midnight Sun 7
How to Seduce Apophis 8
How to Infect Mars 9
How to Circumnavigate Mars 10
How to Tidy the Asteroids 11
How to Classify Ceres......................... 12
How to Question Asteroid 16 Psyche 13
How to Tease Jupiter 14
How to Fall for Io............................ 15
How to Colonize Ganymede 16
How to Surprise Saturn........................ 17
How to Advertise Titan........................ 18
How to Overlook Uranus 19
How to Fly By Neptune......................... 20
How to Speak to Pluto......................... 21
How to Fault Eris............................. 22
How to Infer Another Planet 23
How to Grade Comets 24
How to Collect Eclipses 25
How to Defy the Sun........................... 26

PART II OUR NEIGHBORHOOD: THE MILKY WAY

How to Be a Star.............................. 29
How to Stop Being a Star 30
How to Bless the Interstellar Medium 31
How to Mislay Constellations 32
How to Call Proxima Centauri 33
How to Imagine Barnard's Star b............... 34

How to Honor Sirius 35
How to Weave the Stars 36
How to Reminisce with Vega 37
How to Feed Trappist-1 38
How to Appoint the North Star 39
How to Blush Like Betelgeuse 40
How to Recycle Rigel 41
How to Notice a Dark Nebula 42
How to Observe the Horsehead Nebula 43
How to Mourn Kepler's Supernova 44
How to Nap Like PSR J1841-0500 45
How to Weigh Yourself 46
How to Hide the Milky Way 47
How to Betray Sagittarius A* 48

PART III OTHER AND FURTHER

How to Sail the KBC Void 51
How to Enlist in Andromeda 52
How to Glimpse the Icarus Star 53
How to Lie About SN 2213-1745 54
How to Construct the Elements 55
How to Count Astronomically 56
How to Trick a Trickster 57
How to Scale the Cosmos 58
How to Build a Ladder 59
How to Fathom a Light-Year 61
How to Commemorate the Big Bang 62
How to Perplex Fermi 63
How to Refute Relativity 64
How to Forfeit the Future 65
How to Overlook Differences 66
How to Curve Spacetime 67
How to Think of Time 68
How to Seek Your Roots 69
How to Reclaim Levity 70
How to Designate the Stars 71

How to Battle Entropy........................72
How to Practice Solipsism 73
How to Stay Young 74
How to Drink Water 75
How to Emulate Black Holes 76
How to Endure...............................77
How to Survive a Black Hole 78
How to Mislay 95% of the Universe 79
How to Dance with Dark Matter................80
How to Brand Dark Energy 81
How to Alphabetize the Universe 82
How to Unify 83
How to Calculate the Universe 84
How to Navigate Our Universe 85
How to Play the Lottery 86
How to End the Universe 87

PART IV PIONEERS

How to Invent the Calendar....................91
How to Chronicle the Heavens 92
How to Look Far 93
How to Extinguish Ether 94
How to Illuminate 95
How to Walk from War.......................96
How to Heed Hydrogen 97
How to Hear the Stars........................98
How to Appreciate Technetium.................99
How to Play with Propellants..................100
How to Help Hubble 101
How to Diminish the Space Race 102
How to Merit a Nobel Prize 103
How to Detect Solar Neutrinos 104
How to Spin in Space 105
How to Exit................................106
How to Dishonor Hawking 107
How to Excavate the Past 108

How to Image Exoplanets.....................109
How to Go First...............................110
How to Go Second............................111
How to Go Third 112
How to Go Fourth 113
How to Go Fifth................................114
How to Go Sixth 115
How to Go Seventh 116
How to Go Eighth 117
How to Go Ninth...............................118
How to Go Twelfth 119
How to Go Seventeenth 120
How to Die for Space.........................121

PART V SPACE DUST

Old Astronauts.................................125
The Third Man 126
National Air and Space Museum................127
Félicette, October 18, 1963....................128
[Groundless, that first step] 129
[Saturn encounter—]..........................130
Dear Creator....................................131
The Cats of Mars 132
Copernicus's Cat 133
Dear Time.......................................134
The Headlong Year............................135
The Year's Slow Turning 136
The Red Planet.................................137
Mars As It Might Be 138
Holding You 139

Acknowledgments 141
Original publication 142
About the Author............................ 144
Also by Mary Soon Lee...................... 145

PART I
OUR BACKYARD:
THE SOLAR SYSTEM

How to Paint Mercury

Permit him to pirouette,
spinning as you paint.

Capture his contrasts.
Hammered heat, hidden ice.

Sketch his expression,
capricious, captivating.

Do not erase his scars,
the craters of a survivor.

Remembering his history,
be patient with his moods.

Tell him he's closest
to the Sun's heart.

How to View Venus

Do not come close.
Do not probe beneath
her shroud of cloud.

Take no telescope,
no tripod, no camera.
Step outside. Look up.

Hers the first star
to wake at dusk,
the last to leave.

How to Typecast Venus

Dismiss her as disagreeable
a dyspeptic despot of a planet

preening herself at dawn and dusk
as if she were a star, a deity

when everyone who's anyone knows
she's a desolate acidic dystopia

an abandoned hellhole, her surface
a scorched sulfuric wasteland

which everyone who's anyone knows
is no more than she deserves—

females with pretensions to power
are nothing but shrews or scolds—

never concede, never even consider
the hospitality of her clouds

how their more temperate clime
might harbor an alien life

a life that gladly names her home
that's sheltered by her strength.

How to Colonize Venus

Scorn her scorched surface,
scarred by lava, parched,

hot as hell, as molten lead,
even the pressure punishing.

Above, amid the acid clouds,
Venus approaches hospitable—

her temperature quite temperate,
atmosphere no longer crushing,

oxygen, nitrogen, water available
from the buoyant sulfuric airs—

abide there, thirty miles up,
in drifting aerostatic cities,

an architecture framed in light,
domed, bubbled, suspended in air.

Removed from the mundane realm,
reach for the Word, the Way,

a harmonious merger of human,
machine, artificial gardenscape

where old hurts are healed,
where hatred has no hold,

where rivals hear each other out
or halt, mid-argument, to watch

a hotshot hummingbird show off
its iridescent acrobatics.

How to Thank Earth

Name her animals,
number her rivers,
climb her mountains.

Girdle her with gardens,
gift her with flute and drum,
thank her for your generations.

When you are grown,
when you leave for the stars,
write home.

How to Decorate the Moon

No banners, no balloons,
no gaudy gilded gazebos.

No neon lights, no temples,
no towering triumphal arches.

Instead, on the far side,
an array of radio antennae:

Shielded from Earth's babble,
our ears on the universe.

On the near side, minimalist,
the six Apollo landing sites.

Faded flags and footprints
beneath a canopy of stars.

How to Witness the Midnight Sun

Vacate the equatorial latitudes,
chasing the lengthening days.

Enter the extremities of the Earth,
home to seals, penguins, polar bears.

Look up, there where the summer sun
flaunts itself twenty-four hours long.

Or ignore the preceding instructions.
No need to travel, to make a fuss.

Stay put. Wait for the midnight moon
to offer up the sun's reflection.

How to Seduce Apophis

Open with an invitation
to the little asteroid,

a private appointment
with the planet Earth:

Friday 13, April 2029,
a day ripe for apocalypse.

Raise Apophis' hopes
as you nudge him nearer.

Even little guys dream
of making an impact.

Whisper of the dinosaurs
destroyed by an asteroid.

Toy with him till you tire
then break your promises.

Cancel your invitation.
Let him look, but not touch.

How to Infect Mars

Drop probes upon its rusty soil,
all landers sprinkled with spores
despite sterilization attempts.

Sporadically deploy rovers,
every earnest robot explorer
further spreading contamination.

If such efforts prove insufficient
to overcome inhospitable conditions,
send humans rife with microorganisms.

How to Circumnavigate Mars

Augment atmosphere.
Carve encircling canal.

Punctuate arid plains
with expanses of topiary.

Build timbered barge,
a turf-topped towpath.

Harness your horses,
speak to them gently.

There is no hurry;
walking pace will serve.

Study the unwinding land.
Rock, ridge, horizon.

Remember the pioneers:
Mariner, Viking, Pathfinder.

Sleep on the bare deck
beneath the Milky Way.

No compass, no timetable.
End where you began. Home.

How to Tidy the Asteroids

First circularize their orbits.
The belt is no place for eccentrics.

Next dust each one carefully,
cleaning inside every crater.

Order them by size and disposition:
carbonaceous, metallic, silicate.

Pull the littlest ones in close
where Mars can check on them.

How to Classify Ceres

Comet? No. Planet? No.
Asteroid? Dwarf planet?
Probable protoplanet?

Gravitationally rounded.
Sporadically atmospheric.
Moderately inclined.

Cratered, cryovolcanic,
cradling a solid core
within watery swaddling.

Kind. Tired of seclusion.
Amenable to immigrants,
company, conversation.

How to Question Asteroid 16 Psyche

Do not ask who did this to her,
who shattered her to her core.

Do not ask what she remembers
of the larger self she's lost.

Examine her battered surface,
drenched in frozen tears,

tears shed by lesser asteroids
to cloak her naked metal.

Ask only if she's comfortable,
offering to move her sunward.

Then be silent. Keep her company.
Let her speak in her own time.

How to Tease Jupiter

Pretend you only dropped by
for a chat with Callisto.

Gossip about her sister moons
until Jupiter interrupts.

If he asks how he's looking,
keep staring at his red spot.

Tell him to tighten his belts,
that his equator is bulging.

Then ignore his indignation,
his inordinate tempests.

Remark that manners, not mass,
are the measure of a planet.

Before you leave, make amends.
We only tease those we love.

How to Fall for Io

Study Jupiter's ardent ode
to Io—how her lava flowed,

reversing the schemes of traitors
who had branded her with craters.

The sonnets of Europa too
will teach you much you never knew,

how Io circles in sweet sync,
orbit resonant as her wink.

Both moon and planet ply their verse,
pulling her forward and reverse,

a tug of war that makes her burn,
her insides heating in their turn.

Dream it was you who sparked that fire,
your lines that filled her with desire.

How to Colonize Ganymede

Let humans scratch the surface
of this moon larger than Mercury,
planting their flags, posturing.

Beneath his icy surface,
a buried breadth of oceans,
layer on liquid layer.

Save those for the whales.
Let them bring light.
Let them sing.

How to Surprise Saturn

Praise his thrift,
turning ice, rock, dust
to homespun necklaces.

He'll thank you, then wait,
with resigned dignity, for you
to rhapsodize about those rings.

Do not. Look at him instead.
Admire his supple shape,
his pastel palette.

How to Advertise Titan

No need to mislead; no need to imply
that Saturn's rings nobly straddle our sky.

No need to gloss over the cold dark days,
every view shrouded in dull orange haze.

No need to promote the pristine terrain
as if all our lakes weren't frigid methane.

No need to highlight low radiation,
or hide the cost of a short vacation.

Sufficient to know that here we can fly,
riding the heights of the dim orange sky.

For gravity's grasp's too slight to prevent
our flight through thick air, our golden ascent.

Nothing else matters besides our wings;
soaring in glory, like eagles, like kings.

How to Overlook Uranus

Ignore those lacking labels
with a superlative seal.

Those neither first nor last,
neither largest nor least.

Ignore those on the periphery
who take a sideways view.

Those who speak up rarely,
weighing each soft word.

How to Fly By Neptune

Do not taste his magnetosphere,
his tilted manipulations
of hapless particles.

Do not behold his dusty rings,
their shepherd moons,
their dark arcs.

Do not descend close enough
to sense the inner heat
beneath his storms.

Remember the gravity of his crime,
why the Sun sentenced him
to this far orbit.

Do not listen to his pleas.
Stop up your ears.
Be resolute.

How to Speak to Pluto

Address him as Sir or Pluto,
never by minor-planet designation.

Avoid the terms planet or dwarf;
he is more than his classification.

Ask after his moons, their moods,
his long slow dance with Neptune.

Listen more than you speak,
respecting his age, his orbits.

His views on the Kuiper Belt
are eccentric, but worth hearing.

If he should question you,
answer without trepidation:

Pluto does not gossip.
He will keep his promises.

How to Fault Eris

For refusing to revolve in even
an approximation of a circle.

For skulking off the ecliptic,
spurning accepted inclinations,

furtively avoiding discovery
by twentieth-century astronomers.

For being too big for her boots,
conniving for planetary status,

thereby upsetting Pluto's place,
sowing discord in the system.

As though outweighing him
would ever dwarf his stature.

How to Infer Another Planet

Study the orbital patterns
of trans-Neptunian objects

sentenced to the far reaches,
the cold a constant penance

any hope of pardon or parole
frail as the Sun's pale light.

Notice Neptune nudging them,
bullying them where he beckons

then find that smaller fraction
freed from his interference

sift through their parameters
searching for common cause:

a planet in the outer limits
far farther than demoted Pluto

a planet doing what it can
to help those who have least

the Sun's last sentinel,
standing watch for trouble.

How to Grade Comets

Do not grade on a curve.
Do not use numbers at all.

Even the smallest comet
deserves more than a score.

Consider their constancy,
their patience, their heart.

That long cold trajectory
before the sunward curl.

Compliment their commitment,
their comae, their tails.

The courage it takes to burn
for the sake of their art.

Grant each a gold star
to welcome them back.

How to Collect Eclipses

Select as your first quarry
a commonplace partial lunar.

Measure the Moon's motion,
the seasons of syzygy.

Pick the hunting ground.
Stay as still as you can.

Even these lesser lunars
will seek cover in clouds.

Once the prey is sighted,
let the camera capture it.

There. Savor your triumph,
that spreadeagled image.

Progress from partial lunars
through full blood moons.

Catalogue each new specimen,
itemizing time and type.

Stalk the solitary solars,
the elusive selenelions.

Beware blindness, obsession,
the occult call of shadows.

Earth cannot calm such fervor,
cannot quench fevered fixation.

You will yearn for other skies,
the far transits of Jupiter.

How to Defy the Sun

Not by sailing the seven seas,
nor by soaring the solar wind.

Not by building taller towers,
nor stretching for the stars.

Not by seeking to outshine him,
nor by studying his reactions.

Not by mastering molten metal,
nor by measuring the atom.

All such acts he will applaud,
he who burns to give us life.

Rather, to defy him, despair.
Retreat. Let darkness fall.

PART 11
OUR NEIGHBORHOOD:
THE MILKY WAY

How to Be a Star

Gravitationally collapse a nebula.
Fuse hydrogen into helium.
If desired, explode.

How to Stop Being a Star

Eschew explosions, the egotism
of those who would be supernovas.

Fuel spent, calmly cease fusion,
relinquishing stellar status.

Let your last starlight linger
while you languidly cool down.

A fall from such a pinnacle
unfolds in stately fashion—

not the meager measure of time
the universe has yet existed—

but rather a far greater expanse,
a million billion years or more.

Do not count the hours, the days,
fading degree by slow degree.

When you have set aside the past,
dim to the shade of the backdrop.

A black dwarf, blaze blown out.
Unseen. Unseeable. Unknown.

How to Bless the Interstellar Medium

Burn incense on a twilit beach,
waiting for the sea of night,

or offer fruit, flowers, flame,
flamingos in flamboyant flight.

Praise its dust, gases, clouds;
proffer probes, prostrate yourself.

Send psalms across the light-years
to where it births the stars.

Chant the names of its molecules,
settlers of its lonesome expanse:

hydrogen, hydrogen, hydrogen,
methylidyne, hydroxyl, water,

hydrogen, hydrogen, hydrogen,
methanol, ammonia, formaldehyde,

hydrogen, hydrogen, hydrogen,
cyanogen, cyanoacetylene, cyanamide,

hydrogen, hydrogen, hydrogen,
glycolaldehyde, aminoacetonitrile,

hydrogen, hydrogen, hydrogen,
and the building blocks of life.

Your flesh fashioned from its dust,
your breath the stuff of stars.

How to Mislay Constellations

Set them down carefully,
each star in its place,

artfully arrayed against
a bare black backdrop.

Undertake a brief study
of a mere million years:

an initial investigation
of Saturn's interior,

or a preliminary assessment
of Mercury's mood swings.

After writing up the results,
survey the celestial sphere—

your constellations in pieces,
images mangled and reframed,

the miscreant stars skating
the dark sky to new friends.

How to Call Proxima Centauri

Hail him by ham radio,
aiming the transmission
out toward Centaurus.

Settle down comfortably.
An eight-and-a-half-year lag
slows conversation.

So be neighborly,
but not impatient.
Thank him for dropping by.

Be tolerant of his temper,
the sudden flare-ups,
the awkward silences.

For the next 25,000 years,
he's the closest we'll get
to companionship.

How to Imagine Barnard's Star b

Imagine the star first,
cast against Ophiuchus.

Such a small red dimness
it hides from our sight.

Imagine Barnard's Star b
huddled close to that sun.

A dark, cold, heavy world,
hoarding her scant heat.

Imagine our far descendants
carving cities in the ice.

Furred, thickset exohumans
fitted to the territory.

Imagine them gathering,
once every shortened year.

The huffing of their breath,
outside in the snowy dark.

Hand in hand, staring up
at where they came from.

How to Honor Sirius

Stand on the clear heights
of a southern mountain.

Track the Dog Star's path
across the nighttime sky.

Wait. Let the waking sun
wash out the lesser stars.

When all the others fade,
watch Sirius wink at you.

A star so bright he glows
straight through the dawn.

Kneel down, eyes on him.
Recite his myriad names.

How to Weave the Stars

Begin before Confucius,
before China's first emperor

thread the stars in tales,
in storied constellations

casting Altair as a cowherd
uplifted to Heaven's height

a star-crossed lover stranded
from sweet Vega by a river

longing longing longing
for their yearly reunion

when magpies, moved to mercy,
form a bridge of wings—

that single night together
before another year apart—

weave a thousand variations
on their meeting and parting

celebrate it in festivals
millennium after millennium

then let mercy move you too,
let them slip their posts

let them leave hand in hand,
unwatched, unsung.

How to Reminisce with Vega

No need to rush him.
Wait for the star to near us
in a quarter million years.

Start with small talk.
Compliment his complexion,
that faint blue tint.

Ask about his associates,
Castor, Fomalhaut, et al,
how far they go back.

For matters more substantial,
let Vega take the lead.
Any breakup is hard—

even just a rogue comet
on his system outskirts,
a careless collision—

just a rogue comet,
a commonplace calamity,
but look at the damage—

that disk of debris,
rock splintered to dust
in cascading destruction—

Don't hound Vega. Don't demand.
In time, he may tell tall tales
of his favorites, now gone.

Vega spinning so fast
he's bloated, oblate,
almost coming apart.

How to Feed Trappist-1

Swim the silver river of the Milky Way
to where Aquarius pours his waters out.

Quiet down. Listen for Trappist-1
murmuring to her brood of planets.

So small a star to mother so many,
so intent on her responsibilities.

Every planet tucked safe in place,
rolling round in ordered harmony.

When you've had your fill of watching,
scatter good wishes like breadcrumbs.

How to Appoint the North Star

Search the celestial sphere
for that most constant star
round which the others wheel.

Ordain it with due ceremony.
Inscribe its name in the log.
Bid sailors steer by it.

When the North Star wearies,
let it wander. Name another.
Such service takes a toll.

How to Blush Like Betelgeuse

Summon shame, humiliation,
your most embarrassing moment.

That serves as a beginning,
but cannot match the big guy.

Unbelt. Bloat yourself
until you stretch past Mars.

Shout loud your failings,
every devastating detail.

There, that reddened glow,
so fierce you will explode.

How to Recycle Rigel

Leash the youngster to Orion's foot.
Leave him to burn his fury out—

tearing through hydrogen
in fast, ferocious fusion—

falling in hunger upon helium,
carbon, neon, oxygen, silicon—

anything to fan the flames,
anything to get a reaction—

antics predictably culminating
in catastrophic collapse.

Sweep up his cindered remains;
reuse in calmer stars.

Not all sinners can be saved,
not all wild hearts tamed.

How to Notice a Dark Nebula

Disregard the galaxy's glare,
the cocky constellations.

Search for a smudged dimness
dusking the brighter sights.

That dirty interstellar dust
disguises secret treasure.

Within her curtained darkness
the birthing stars lie swaddled.

Stay as still as you can.
Do not interrupt her lullaby.

How to Observe the Horsehead Nebula

Do not replay the same dark horse,
the same perspective others dealt.

Aim Hubble's wide field camera
near Alnitak, Orion's Belt:

the old made new in infrared,
the horse's head now dusky rose

a prima donna girt with stars
preening in picture-perfect pose.

Or shut your eyes and journey back
to Boston, eighteen eighty-eight

where Williamina Fleming
peers at a photographic plate—

tired, perhaps, or missing Scotland,
dreaming thoughts above her station,

abandoned wife, single mother—
first to see that indentation.

How to Mourn Kepler's Supernova

Armor yourself with amulets.
Go barefoot under the night.

Scan the stars north of Scorpius
where Ophiuchus grasps his snake.

Buried between those markers
dire remnants of destruction.

The scars too faint, too far
for any eye to pick them out.

Yet back in Kepler's prime,
they were a fiery funeral pyre.

An outpouring of grief so bright
it burned through daylit sky.

Look up therefore and weep
for what was rent asunder.

Call on your gods to ease
the unbinding of a star.

How to Nap Like PSR J1841-0500

Cease your frenzied signaling,
that spinning lighthouse beam.

Even the most stalwart stars
deserve respite from strife.

Nestle in the curving arm
of the Milky Way's embrace.

Drowse for a year, dreaming
of a prince's wakening kiss.

Enough. No prince will come.
Wake up. Resume your labors.

How to Weigh Yourself

Not on the bathroom scales,
not in ounces nor in grams.

Consider your protons,
so small yet positive.

Consider how you outweigh
each constituent proton

by more than ten thousand
trillion trillion times.

Consider how the Milky Way
in his turn outweighs you

by more than ten thousand
trillion trillion trillion times.

Between those two extremes,
consider yourself, your works.

Consider the breadth of time:
billions of years elapsed,

billions left before the Sun
burns out. All things end.

Yet every act, every decision
inscribed immutably in spacetime.

No kindness too small to mark.
Reach out to those nearby.

How to Hide the Milky Way

Sunlight is simplest,
one star overruling the rest.

Between dusk and dawn,
customary methods serve:

fire, fog, fumes, full moon
to flood the river of stars.

Or stay within city limits,
shielded by electric glare.

Avoid wilderness, desert,
the star-strewn ocean.

If all else fails, look down.
Heed not his siren call.

How to Betray Sagittarius A*

Scorn her screams,
those wailing radio waves

warning of the hole
in the galaxy's heart.

Worthless to waste grief
on ancient tragedies.

Repudiate her testimony
of mass assassination,

her lone voice speaking
for the silenced stars.

PART III
OTHER AND FURTHER

How to Sail the KBC Void

Quit the comparative clutter
of the universe's structures,

filaments thick with galaxies,
clusters in constant clamor.

Beyond those crowded walls,
the last great open spaces.

Step aboard the Milky Way.
Voyage the void's vastness,

unmoored amid an ocean
a billion light-years deep.

Wave at the nearby ships:
Andromeda, Triangulum, M32.

No matter how it seems,
we are none of us alone.

How to Enlist in Andromeda

Complete the requisite forms:
name, age, mass, velocity,

preferred branch of service,
your residential address

in a minor spiral arm
of a neighboring galaxy.

Renounce prior allegiances
in favor of Andromeda,

pledging your willingness
to support and defend her.

Wait four billion years
to report for duty.

How to Glimpse the Icarus Star

Cast Hubble high into the sky.
Peer through his mirrored eye.

Let gravity augment his sight,
massed galaxies lensing light.

Across nine billion years of night
the image of a star takes flight.

A star, a single blue-hot star,
from almost unimaginably far.

Nicknamed Icarus for a boy
whose fame outlived his joy.

Hubble's portrait so very old,
the star itself long dead and cold.

How to Lie About SN 2213-1745

Deny she was ever a star,
or, failing that,
deny she ever went nova.

Deny that she burnt
brighter than galaxies,
brilliant in ultraviolet.

Or admit those facts,
ten billion years old,
and invent the backstory.

Pretend she was spurned,
that despair drove her
to destruction.

Deny her strength.
Deny her independence.
That she chose to shine.

How to Construct the Elements

Ingredients:
one expanding universe,
0.01 seconds old.

Handle with care:
approximate starting temperature
a hundred billion degrees.

Stir stew of matter
for three or four minutes
until deuterium formation.

Bake primordial hydrogen
in stellar furnaces,
concocting heavier nuclei.

To advance beyond nickel,
collapse massive stars
as Type II supernovae.

Season to taste,
scatter throughout universe,
serving up in planets, stardust.

How to Count Astronomically

Learn in the stellar nursery
the integrity of integers.

One moon to circle us;
one sun to circle round.

Declination, right ascension:
two coordinates on a sphere.

Three, the third dimension,
to stipulate the distance.

Four terrestrial inner planets:
Mercury, Venus, Earth, and Mars.

Five fundamental forces,
four overt, the fifth covert.

Learn of leptons and of quarks,
the elementals of the stars.

One, two, three: up, down, strange;
four, five, six: charm, bottom, top.

Then memorize the speed of light,
299,792,458 meters per second.

When you leave the nursery,
set aside your certainties.

Our universe approximately
fourteen billion years old,

the precise age a secret lost
at the far origin of time,

redshifted, forever receding;
truth approached, never reached.

How to Trick a Trickster

Look for a loophole,
a ruse, a lure to wager
in place of your soul.

Offer the canals of Mars
or the moons of Venus
or the ice of stars.

Or venture such things
as Saturn's mountains
and Mercury's rings.

Pledge the sum of gold
that the universe owned
when it was one hour old.

And then to follow
add the women launched
by Project Apollo.

How to Scale the Cosmos

Begin with the basics:
quarks, leptons, bosons.

Minuscule building blocks
to assemble atoms, molecules.

Each hydrogen atom fitted
in a ten billionth of a meter.

Fashion stars from hydrogen
in assorted shades and sizes.

Dwarves of modest diameter,
a hundred thousand miles.

Sprawling supergiants outstretched
billions of miles from side to side.

A trillion trillion stars astride a stage
a hundred billion light-years wide.

Scatter them in galaxies.
Sprinkle with dust and darkness.

How to Build a Ladder

Not of wood, nor of metal,
nothing small, nothing simple.

Don't try to reach the treetops,
but the farthest trembling stars.

Not a ladder that you clamber
in steel-toe boots and harness

but a ladder built of light
to plumb the deeps of space,

each rung linked to the last
in ever expanding scale,

from the first small step
of lunar laser ranging,

measuring the Moon's ambit
with meticulous precision,

then up and up by parallax,
stellar and spectroscopic,

determining the distance
of denizens of the Milky Way,

then up and up where Leavitt
led us with her candlelight,

spying Cepheid variables to put
the galaxies in their places,

(continued)

then up to sound the firmament
by the flare of supernovae,

up to the outermost reaches
revealed through Hubble's law.

Contemplate that absurd immensity,
ten billion light-years or more,

then come back down to Earth,
to household chores, a cup of tea.

How to Fathom a Light-Year

Imagine walking for a century,
never resting, never veering,

the tread of your progress
unwrapping a ribbon behind you,

each year's unrolling ribbon
sufficient to girdle Earth.

Imagine stopping at the last,
looking back along your track:

an immensity that light crosses
in a quarter of a minute,

how much vaster the enormity
that light must measure in a year.

Too much to grasp alone.
So begin again. Ask for aid.

Imagine the whole of humanity
joined in shared pilgrimage,

the account of our steps
added, accumulated, merged,

how together we might cover
a light-year in ten days,

walking side by side, singing,
the hale helping the halt.

How to Commemorate the Big Bang

Apprehend all you can
of her singular nature,
the rumors of our origin.

Wonder at her works,
the weight and light and height
of the universe she wrought.

Observe an eon's silence,
hand over heart,
hearing the echo of her passing:

that relic radiation
written in microwaves,
fading into eternity.

How to Perplex Fermi

Direct radio transmissions
anywhere else but Earth.
Steer starships clear.

So many billions of stars,
so many planets teeming
with tentacled life.

Convene at Canopus for cocktails
while Professor Fermi ponders
your paradoxical absence.

How to Refute Relativity

Rely on common sense,
not Einstein's equations.

All clocks, if accurate,
tick time at the same rate.

All yardsticks, moving or not,
must still measure a yard.

Child's play to know falling
is not equivalent to floating.

Why if you thought that,
you'd think space was bent!

And how can simultaneity
be a matter of perception?

Other observers are not equal.
Your truth, the only truth.

How to Forfeit the Future

Never search to circumvent
the light-speed restriction.

Never labor to load the dice
at the quantum level.

Never dismantle stellar systems,
running a race against entropy.

Never colonize the galaxy;
never build a single starship.

Never strive. Never explore.
Never venture beyond the Moon.

Moderate your ambitions.
Stay home. Refuse the stars.

How to Overlook Differences

Remind yourself that everything you see—
skyscrapers, sidewalks, soda cans,
pylons, power lines, pine trees,
pigeons, spiders, a cantankerous drunk,
a cackle of teenagers, poor people, rich people,
young, old, immigrants, idiots, ideologues,
introverts, extroverts,
the last person you argued with—
that all of them are traveling
through spacetime
at the speed of light,
as are you,
all of us rushing so fast
even when we feel we're going nowhere,
all of us headed in pretty much
the same direction.

How to Curve Spacetime

Construct a universe.
Smooth its fundamentals
in a semi-Riemannian manifold
where mass and energy shape spacetime,
a universe formed on symmetries,
curved by star-matter,
gravity grounded
in geometry.

How to Think of Time

Howsoever pleases you—
as a river, an ocean,
a fatherly figure,

finite or infinite,
linear or looping,
forgetful or forgiving.

Howsoever you please,
but if you're perplexed
imagine an immanence:

an immaterial immanence
inextricably entangled
in curvaceous spacetime

with surplus dimensions
(some six or seven to spare)
compactly curled up.

Brace yourself. Go further.
Imagine an imaginary time
orthogonal to ours.

Imagine its imaginary history
as an imperfect sphere,
slightly squished.

Imagine our reality
determined by summing
myriad imaginary histories,

each history replete
with its own grievances
to forgive or forget.

How to Seek Your Roots

Genealogy, anthropology, paleontology
offer only a cramped perspective.

To find your remotest origins, search
before the Earth itself was formed

before Earth's raw ingredients
were assembled by the stars

before taxes, before art, before words,
before flowers, or pebbles, or planets.

Search not for what makes you unique
but for what we held in common

before deuterium and helium emerged
in the first minutes of the universe

back thirteen billion years and more
to the first fraction of a second

before protons, before neutrons,
before gravity parted from unity

before the first symmetry broke,
before any inkling of grief.

How to Reclaim Levity

Relinquish relativity, relinquish Newton
for Aristotle's centered universe

the world midmost on her still throne
beneath the nested celestial spheres

twin powers summoning the elements
to their right and natural places

gravity to draw down earth and water
and levity to uplift air and fire.

Praise that partnership's lost half,
the rising counterpoint to gravity.

Praise every frivolous foolishness
that holds us up when we are down.

How to Designate the Stars

Assign them the numbers
of their celestial coordinates,
declination, right ascension, date,
as if a sphere spun round the Earth
with all the flickering stars
pinioned in place.

Now put aside your instruments,
arm yourself with a sky chart.
Pick out the constellations,
befriend their marker stars.
Call them by their names.
Let them speak to you.

How to Battle Entropy

Perpetual motion machines.
Cosmological constant adjustments.

Increase improbability.
Reverse the irreversible.

Write large your defiance
in the annals of spacetime.

Fight though the stars falter.
Fight though you fall.

How to Practice Solipsism

Forgo philosophy for physics,
forsaking Gorgias, Descartes

the purely hypothetical fancy
that non-self may be nonsense.

Found your position on fact,
the fundamentality of numbers.

Against the unforgiving forever
of a universe infinite in time

thermodynamic equilibrium
is subject to fluctuations:

uncounted combinations flicker
into form before degenerating

random aberrations summoning
atoms or minarets or minds

the Taj Mahal, the rings of Saturn,
a scruffy tabby licking her paw.

Amid such seas of possibility,
the laws of probability dictate

it's far more common to shape
the small, the modest, the singular

and so you're likely just a brain,
a bodiless brain all by itself

imagining for a fleeting instant
that there is more to reality.

How to Stay Young

For merely minor adjustments
move down to the Earth's core.

Deep in her gravitational well,
time ticks more nonchalantly—

shaving a couple of years or so
from each four million millennia.

For more major rearrangements
accelerate to near light speed.

Take in a little sightseeing
while nations rise and fall,

a solution effective enough
yet lacking a certain gravitas.

For those who would preserve
both youth and decorum,

approach the event horizon
of your favorite black hole.

Settle down with a good book
as the universe ages around you.

How to Drink Water

Gratefully, mindfully,
remembering its source,
reservoirs, rivers, rain,
a gift of the clouds,
and before that,
further and further,
how hydrogen formed
in the hot aftermath
of the Big Bang,
first singular protons,
then their pairing
with electrons,
then the mustering of hosts
of hydrogen and helium
in stellar furnaces
where fusion's assembly line
marshalled heavier atoms,
resulting, among others,
in oxygen, atomic number eight,
how oxygen volunteered
to join hydrogen
in water molecules,
how that water infiltrated
the gas and dust that made
our Sun, our solar system,
our Earth of dirt,
sea, streams, springs,
the enormity of that tour of duty
before ever you thought
to pour it in a glass
or cup it in your hand
to drink.

How to Emulate Black Holes

Dress in dull, dark drabness,
forgoing frippery, flirtation.

Frivolity forsaken, accept all
who are drawn by your gravity.

Do not refuse the homeless,
the helpless, the lonely.

Never betray their secrets.
Never let go of those you love.

How to Endure

Arise from ashes, from cataclysm,
the collapsing of a star's core.

Enter the brethren of black holes,
bonded in common determination.

Let no particle, no photon leave
the bounds of your dark halls.

Do not despair. Do not surrender.
Hold steady through the eons.

Outlast the firmament of stars,
staying when the lights go out.

Inheriting a darkened universe,
stand guard with your brothers.

Your massive presence a monument
testifying to the luminous past.

How to Survive a Black Hole

Have patience. Outlast it.
Time is the enemy of kings,
evaporating all things
in ten-to-the-hundred years.

Time is the enemy of kings.
Wait till the universe cools.
In ten-to-the-hundred years:
only photons and their peers.

Wait till the universe cools—
how are the mighty fallen!
Only photons and their peers;
no more strife, no more fears.

How are the mighty fallen?
By evaporating à la Hawking.
No more strife, no more fears.
So have patience. Outlast it.

How to Mislay 95% of the Universe

Add up the stars,
their attendant planets,
moons, comets, asteroids.

Append sundry items:
dust, gas, black holes,
a nuisance of neutrinos.

Then stop. Close up shop.
If some mass is missing,
assume it doesn't matter.

Do not inquire
what holds the galaxies
from coming asunder.

Nor ever pause to think
what dark energies
might drive us all.

How to Dance with Dark Matter

Close your eyes. Let her lead,
she who steers the stars.

No need to fret about form,
fashion, the cut of your suit.

No need to fret about the steps,
or the right thing to say.

Her unseen strands the chords
that conduct the cosmos,

orchestrating the structure
of galactic superclusters.

Listen to the music of her gravity.
She will let you lean on her.

How to Brand Dark Energy

—after a remark by Brian Clegg

One can hardly be expected
to refer to it
as that ineffable entity
underpinning the apparent accelerating expansion of the
                                                  universe—

so pin a label to it—
call it dark energy—
as if we knew it to be dark,
as if we knew it to be energy—

perhaps it would be better
to name it as if it were a pet—
Mister Floofy, maybe, or Bitsy—
some friendly unintimidating descriptor
to cover up the fact
that the universe may be coming apart
at the seams.

How to Alphabetize the Universe

Aberration, achondrite, albedo,
aperture, aphelion, apogee, Apollo—
apologize for approaching apocalypse—
Big Bang, Big Crunch, Big Rip, black holes.

Carefully place celestial spheres
after their celestial poles.
Demand dark matter defer to dark energy,
that eccentricities precede eclipses.

File Ganymede under Galilean moon.
Hurry to index Jupiter, the Kuiper belt,
Lagrange points, Mars, Neptune, Oberon,
once-planetary Pluto, quarks, quasars.

Reflect on relativity's relevance,
that spots sully even the Sun,
that Titan is distinct from Titania,
that umbral material may coat Umbriel.

Move the Valles Marineris before Venus,
waning before waxing, X-class before X-ray.
Shelve years by duration (Mercury first),
zeroing in on the Zen of the zodiac.

How to Unify

Find in the apple's fall
the selfsame force that moves
the planets in their courses.

Find in the magnet's pull
the selfsame electricity
that governs light waves.

Find the laws that bind
space to time, that marry
matter to energy.

Stop. Rest. Then broker peace
between those old adversaries,
gravity and the quantum.

How to Calculate the Universe

Begin with tally marks scratched
in clay or bone or stone

carving days, full moons, horses,
the year's patterned mysteries.

Then claim numerals, fractions,
zero, all that is rational—

systems sufficient for cities,
tithes, irrigation, gardens.

Applying an analytic attitude
lay down the lines of latitude

each new horizon yielding
the glory of a wider view.

Measure the multitudinous stars
by magnitude and parallax.

Extract with equal exactitude
the dimensions of atoms

harnessing fresh inquiries
to each hard-won certitude.

When your questions exceed
the reach of integrated circuits

fashion a quantum computer:
open the halls of Hilbert space.

Step inside. The universe is vast
and full of wonders to uncover.

How to Navigate Our Universe

Not by compass or polar star,
neither able to steer you far.

Not by fickle constellations,
adrift without foundations.

Not by sight, not by any light,
when you pilot the sea of night.

Hold hard the helm and for your guide
take the tug of gravity's tide.

Left by matter, seen and unseen,
dark energy its hidden queen.

The warp and weft that fold the sky,
bending spacetime as you sail by.

How to Play the Lottery

Don't settle for pedestrian prizes,
payouts of a paltry million dollars.

Play with the highest of high rollers.
Pull up a chair beside Death's throne.

Play for an astonishingly improbable,
all-encompassing, astronomical prize—

play for vacuum decay, that sudden shift,
the Higgs field realizing its potential—

as a ball balanced precariously plummets
to a preferred, lower resting place—

settling comfortably in a new spot
in the cosmic casino's roulette wheel—

though in this case the ball might shift
due to spontaneous quantum tunneling—

and the ball rolls at the speed of light
arriving before you see its approach—

and without the ball, which is not a ball,
there would be no mass, no weight at all—

a prize to share with the whole world,
spreading unstoppably every which way—

a prize beyond diamonds and gold,
one to unbind the laws of chemistry—

unmaking you, me, Earth, stars, galaxies.
Play to lose. Play to lose. Play to lose.

How to End the Universe

1. Big Crunch

Crumple it up,
smaller and smaller,
a mote of matter. Begin again?

2. Big Rip

Bring rage to bear,
tear it apart, shredding stars,
the fabric of spacetime.

3. Big Freeze

Turn down the lights,
dismiss the dimming stars,
sweep the bare stage.

PART IV
PIONEERS

How to Invent the Calendar

Begin before Babylon,
before Stonehenge hefted
from Salisbury Plain.

Measure the Moon's moods,
its waxing and waning,
the shortening days.

Dig a dozen pits,
each one to designate
a fickle lunar month.

The months mismatched
to the cycling seasons,
drifting out of date.

Anchor their count
by the solstice sun
notched in the hills.

A Scottish calendar
carved in a curving arc
before Scotland's birth.

If this does not suffice,
step still further back,
mark time on cave or bone.

How to Chronicle the Heavens

Before the sun lifts
from the eastern horizon
search the twilight sky.

Against that soft dawning
find the first faint sight
of Venus returning.

Or watch the westering sun
for the final glimpse
of Venus departing.

Carve a cuneiform catalog
commemorating the planet's
comings and goings.

A truth to outlast empires,
transcribed from the sky,
captured in clay.

How to Look Far

In memory of Galileo Galilei
[February 15, 1564 - January 8, 1642]

Select spherical glass surfaces
in curves convex and concave.

Position paired lenses precisely,
bending light to your bidding.

Peering through this instrument,
map the mountains of the Moon.

Reveal the concealed satellites
of Jupiter, unfixed, restless,

their circling paths disclosing
the ordering of the firmament.

No matter what Ignorance insists,
banish lies. Let there be light.

How to Extinguish Ether

In memory of Albert Michelson and Edward Morley

Begin with belief, with trust
in ether's luminiferous essence.

Assemble an interferometer
to measure its influence:

light rays at right angles
reflecting and returning

disclosing differing speeds
in colored fringe patterns.

Begin with belief, but admit
that you were in error:

light travels constantly
no matter the direction

ether's existence erased
by experimental evidence.

Do not sulk. Do not grieve.
Not all failures are futile.

How to Illuminate

In memory of Henrietta Swan Leavitt
[July 4, 1868 - December 12, 1921]

Softly, gently, a candle gleam
outshone by bolder lighthouse beam

take your place in Pickering's horde:
female helpers he could afford

to carry out astronomy
with laudable economy.

Softly, meticulously note
the facts on Cepheids remote.

Softly ponder and then conclude
period scales with magnitude.

Softly, and yet until your light
we had no measure of the night.

How to Walk from War

In memory of Karl Schwarzschild
[October 9, 1873 - May 11, 1916]

Walk from war, its roar of gunfire,
its field equations of ammunition.

Walk slantwise from out its winter
into the expanse of mathematics

where Einstein cleared the way,
that clear pure view, unstained.

Waste no time; you have so little.
Climb the foothills of relativity.

Up. Further. Up untrodden slopes,
past any path to top the peak:

the first exact and perfect solution
of general relativity's field equations.

No pain on those heights, no din,
only a brief solitary splendor

before you must trek back down
to the mess and massacre of war.

How to Heed Hydrogen

In memory of Cecilia Payne-Gaposchkin
[May 10, 1900 - December 7, 1979]

See what others do not see,
the underestimated multitude,
that stars are largely made
of hydrogen and helium.

The underestimated multitude—
that well-known lot of women,
of hydrogen and helium—
defy it, spark a fire.

The well-known lot of women
that stars are largely male.
Defy it. Spark a fire.
See what others do not see.

How to Hear the Stars

In memory of Karl Guthe Jansky
[October 22, 1905 - February 14, 1950]

At Bell Labs' invitation,
investigate interference.

Build a brass antenna,
a hundred feet broad.

Study the radio static,
listening to storms.

Behind their thunder,
a background hissing:

rising and falling
with the sidereal day.

At the galaxy's heart,
an orchestra of stars.

How to Appreciate Technetium

Notice its significant absence,
that gap in the periodic table.

To alleviate the lack of it,
craft the element yourself:

accelerate minuscule missiles
to bombard molybdenum.

Welcome your new creations
without rules or reservations,

cherishing each atom despite
their inherent instability,

as you should later cherish
their discovery in red giants—

a short-lived presence written
in the spectra of the stars,

proof and epitaph to others
born in that nuclear fire.

How to Play with Propellants

Step back in time: before Apollo,
before Gagarin, before Sputnik,

before computers fit in phones,
before software tamed chemistry,

back in the incendiary heyday
of the hunt for hypergols,

when rocket scientists sported
with red fuming nitric acid,

with hydrazine, aniline, pyridine,
amines, aromatics, mercaptans,

with toxic and mephitic fluids
of distinctly vicious disposition—

liquid orthohydrogen stealthily
transforming itself, vaporizing—

ClF_3 igniting water, sand, asbestos,
promising performance at a price.

Abandon caution, commonsense
that you might master fire.

Light a pyre beneath Saturn V.
Thank Prometheus for his gift.

How to Help Hubble

In memory of Edwin Hubble
[November 20, 1889 - September 28, 1953]

Fetch him offerings
of mice, lizards, birds,
a dragonfly for his amusement.

Sprawl atop his papers
to prevent him pondering
pointless extraplanetary topics.

Sleep at the foot of his bed,
except that one last time
when you curl beside him.

After he has receded,
too far, too fast to follow,
watch for him at the window.

How to Diminish the Space Race

*In memory of Katherine Coleman Goble Johnson
and all her sister mathematicians*

Preemptively limit leading roles
to men cut in a military mold,

white men, of course, Ed Dwight
not quite the right stuff.

If circumstances necessitate
employing women or minorities,

leave them from the limelight,
segregated and unremarked.

A woman, even a black woman,
may master orbital mechanics,

may shepherd Alan Shepard
to a safe splashdown,

may double-check the trajectory
of John Glenn's Mercury mission,

women by the hundreds may compute
details essential yet unseen,

but there's no need to admit
men's lives depend on them.

How to Merit a Nobel Prize

—*for Jocelyn Bell Burnell*

Dedication, diligence, discovering
an anomaly in the deluge of data,

an anomaly that you defend
against insistent skepticism,

an anomaly that discloses
existence of the extraordinary:

pulsars, spinning neutron stars,
flashing like lighthouse beams.

If these should prove too paltry
to persuade the prize committee,

if the laurels go to others
as though your part were lesser,

avoid bitterness. Let your worth
be a quiet and shining truth.

And if, four decades later,
acknowledgment should arrive—

a prize for fundamental physics,
a three-million-dollar accolade—

avoid triumph, avoid avarice,
give away your due reward,

opening the door to physics
to those who could not enter.

How to Detect Solar Neutrinos

In memory of Raymond Davis, Jr.
[October 14, 1914 - May 31, 2006]

A mile down the Homestake Mine,
delve for riches rarer than gold.

In darkness, in the hot depths,
search for evidence, a sign:

chlorine transforming to argon
in the alchemy of neutrinos.

Insubstantial, invisible,
unveiled by their actions.

Messengers born in brightness,
forged in the Sun's fire.

How to Spin in Space

Ignore the camera's scrutiny,
ignore every indignity—

deprived of rain, of grass,
deprived even of gravity—

a specimen, an experiment,
orb spider orbiting Earth

in your own time, commence,
extrude silk from spinnerets

amid the floating, flailing
difficulties of freefall

fashion a rudimentary frame,
then radiating strands

such effort to form a web
frayed with messy missteps

rest, gather your resolve,
begin again next day

working alone, in silence,
keeping doubt to yourself

tightrope-walk the threads
to spin a circling spiral

a pinnacle of precision
approaching perfection.

—In 1973, two spiders were taken to Skylab.

How to Exit

Launch from Cape Canaveral,
hard on your brother's heels.

Hurry up and overtake him
somewhere in the asteroid belt.

Complete primary mission:
Jupiter, Saturn, Titan.

Expectations exceeded,
don't try to come back.

Head out of the ecliptic
into the empty heights.

Take a snapshot of home
from that lonely vantage.

Then cross the heliopause
into interstellar space

where none have gone before.
No hesitation. No regrets.

—*Voyager 1 left the solar system in 2012.*

How to Dishonor Hawking

In memory of Stephen Hawking
[January 8, 1942 - March 14, 2018]

Harp on his helplessness,
harnessed to a wheelchair,
his hardships, his handicaps.

Or overpraise him, raise him
on an impossible pedestal,
replace truth with myth.

As if truth were not enough:
that he tried to uncover
how the universe works.

How to Excavate the Past

Leave home, leave Earth,
leave the Sun's warm hearth

to probe Pluto, then pass
beyond the known worlds.

Flung like a stone to skim
the frozen edge of nowhere

for a fleeting encounter
with a planetesimal fossil:

2014 MU69, prototypical resident
of the Kuiper Belt district,

a small, squat, snowman figure.
Steadfast. Stoic. Stubborn.

Still much the same as he was
in the solar system's infancy.

Telling you what we came from
without speaking a word.

—*On 1/1/2019, the New Horizons probe flew by 2014 MU69.*

How to Image Exoplanets

Launch a flock of telescopes
to the second Lagrange point.

Shield them from sunlight;
shield them from earthlight.

Unfurl a starshade for each,
petals opening like flowers,

occluding a distant star
to spy on dimmer planets.

Focus the flock's far gaze
on that revealed light.

The lakes, rivers, forests
of worlds beyond reach.

How to Go First

In memory of Yuri Gagarin
[March 9, 1934 - March 27, 1968]

Light up the darkness
high over Earth's nightside
with your smile, unseen,
on that first orbit.

How to Go Second

Let another take the first step,
let their words resound in history.

Wait nine minutes for your turn
to follow him onto the Moon.

Your own words almost weightless,
an afterthought to Armstrong's.

A magnificent desolation in that,
quarter of a million miles from home.

Take communion alone, a private act.
Let God's grace bless second place.

How to Go Third

Let other countries fight for first:
first in space, first to the Moon.

Wait on the wings. Your turn will come,
Yang Liwei waving flags from orbit.

The march is long. Your turn will come
to claim the Moon, Mars, beyond.

Mining, manufacturing, assembling.
Not who goes first, but who will stay.

How to Go Fourth

Leave monumental ambition
to Roscosmos, NASA, ESA.

Make Mangalyaan's mission
small, simple, economical.

Fourth planet, fourth place,
India's introduction to Mars.

Rank it not by payload mass
but the thunder of applause,

the chromatic crush of saris,
hair garlanded with gajras,

women in the control room
reaching to hug each other.

How to Go Fifth

Refuse the self-importance
of the first four forces—

gravity, electromagnetism,
both the nuclear forces—

all flagrantly flaunting
irrefutable effects.

Remain aloof, elusive,
quintessentially evasive.

Scatter the universe itself
to hide your tracks.

How to Go Sixth

Squeeze between boron and nitrogen,
sixth spot in the periodic table.

Marshal alpha particles by threes
to assemble your nuclei in stars:

two alphas merging as beryllium,
a third combining to make carbon.

Always remember that revelation,
the communion of coming together.

Help others, catalyzing fusion,
bonding with anyone who asks.

How to Go Seventh

In the seventeen minutes before lunar impact
impeccably transmit images of the Moon,
the last sight you'll ever see.

Disregard the naysayers so quick to criticize,
to mock your brothers' misfortune,
discounting their sacrifice—

Rangers I and II constricted to Earth's confines,
coffined high in her atmosphere,
consumed by fire.

Rangers III and V mortified they missed the Moon,
still ceaselessly circling the Sun,
heliocentric, haunted.

Rangers IV and VI, their mission incomplete,
confounded by camera or computer,
crashing moonward—

Do not think of yourself, the coming collision.
Consecrate that final picture
to your brothers.

—Ranger VII crashed into the Moon on July 31, 1964.

How to Go Eighth

Let the other planets
yield their lesser secrets:

Earth succumbing first,
easy prey to explorers.

Venus the second victim,
viewed by Venera, Mariner.

Mars, Jupiter, Mercury,
Saturn scarcely resisting.

Uranus, more reluctant,
holding out for longer.

When they come for you,
dissemble, play dumb.

Pretend Pluto's a planet,
suggest they visit him.

Anything to distract them
from your own dark truths.

How to Go Ninth

At the solar space recital,
vanquish Titan for the title

of best and biggest satellite,
Jupiter's love and his delight.

So many moons—to win first place!
Even in triumph, reach for grace.

First among moons, not among all,
recall that pride precedes a fall.

Bow down to those more vast than you:
the Sun and seven planets too.

Bow last to Mercury, whose worth
exceeds the measure of his girth.

—Ganymede is the ninth largest known celestial body in the solar system.

How to Go Twelfth

—for Valentina Tereshkova

Know your place, a woman's place,
eleven paces behind the men
who lead the way.

Know your duty, know how to shine
without ever diminishing
what they did.

Orbit the Earth forty-eight times,
alone, your stomach queasy,
but that view—

that vast untrammeled revelation—
as unequaled, as uneclipsed
as you yourself.

How to Go Seventeenth

Follow your brother to the frontier,
a pair of explorers, Jupiter-bound.

Rush through scorching radiation belts—
a hundred thousand miles an hour—

skimming close to cloud crests,
snapping pictures of the Red Spot.

Fly by, Jupiter's force flinging you
across the solar system to Saturn—

first to ever reach those rings,
up close, braving their passage,

measuring Saturn's magnetosphere,
taking Titan's wintry temperature.

Don't stop. Don't doubt destiny.
Keep heading onward and outward.

*—Pioneer 11 was the seventeenth space probe
of the Pioneer program.*

How to Die for Space

Bondarenko. Freeman. See, Bassett.
Grissom, White, Chaffee. Komarov.
Williams. Adams. Lawrence.
Dobrovolsky, Patsayev, Volkov.
McAuliffe, McNair, Smith,
Scobee, Onizuka, Resnik, Jarvis.
Vozovikov.
Ramon, Brown, Anderson,
Husband, McCool, Clark, Chawla.
Alsbury.

Before Gagarin's first orbit
or over half a century later

engineer, surgeon, physicist
or a high school teacher

American, Soviet, or Israeli,
Christian, Buddhist, Hindu, Jew

man or woman, black or white,
with their crewmates or alone

breathless, burnt, blunt force
during a mission or in training

not their differences
but what they shared

launching, landing, lost
while reaching for the sky.

PART V
SPACE DUST

Old Astronauts

Brush their teeth each morning,
comb their thinning hair
then read the paper over breakfast,
practiced at ignoring the twinge
when they read a two-paragraph article
about the guys out there,
young men and women
dusting cobwebs off last century's dreams,
that twinge that comes to old folks
who would rather be doing
than reading.

Are endlessly tolerant
of questions, interviews,
of all the strangers, who, even now,
decades later,
see only those handful of days
breathing stars.

Never complain
when the questions
turn back to Armstrong and Aldrin—
those who go first
leave the deepest mark.

Stand in their backyard
after midnight, house lights off
but the show still going on
up there,
stand out in the cold
wondering if Hercules
ever yearned
for a thirteenth Labor.

The Third Man

No matter how crowded the party
he can open a door in his head
to that airless void,
two hundred thousand miles
from home,
as solitary as a man can get,
his face pressed to the glass
of the command module,
looking not for Earth
nor his friends
on the Moon,
but out,
his heart falling
into the infinite.

National Air and Space Museum

In the main hall's towering brightness
they pressed upon me:
Sputnik, Mercury, Apollo, Pioneer,
their honored weight too much to bear.

Retreating to a darkened gallery,
I paused beside a case of items
taken to the Moon:
toothbrush, razor, scissors, floss,
a dose of glory
in manageable proportions.

Félicette, October 18, 1963

Her fifteen-minute flight
a French footnote
in the race to space.

A diminutive tuxedo cat,
conscripted, strapped in place,
streaking through the stratosphere.

No interviews,
no ticker-tape parade
on her safe homecoming.

Euthanized later
for scientific gains,
her small steps stilled.

In memory of Alexei Leonov
[May 30, 1934 - October 11, 2019]

Groundless, that first step
beneath the unblinking stars,
historic spacewalk.

MARY SOON LEE

Saturn encounter—
Mauna Kea telescope—
that luminous glimpse

Dear Creator

Thank you for your submission.
Regrettably, your draft universe
isn't quite right for us.

Perhaps you proposed
a surplus of elements,
or your deterministic rules
were too predictable.

Possibly your Big Bang wasn't,
or you weren't particular
about your particles.

Maybe you treated your material
with insufficient gravity.

We find flat universes
too two-dimensional,
and cellular automata formulaic.

Cosmological constants should be.

Please wait until entropy
increases by at least one percent
before resubmitting.

P.S. The turtle stack amused us,
but we couldn't get to the bottom of it.

The Cats of Mars

Refuse to wear spacesuits,
have never been outside,
never hissed at a dog,
never tasted mouse.

Consume 58.2% of the colony's fish,
most of which they appear
to transmute into fur
to be shed where inconvenient.

Defy gravity (such as it is),
hind legs injecting delta-v
to dock on ductwork
and claw the insulation.

Belong to no nation,
no individual,
crossing borders
as if they own the planet.

Copernicus's Cat

After her master
failed to acknowledge
that the celestial spheres
revolved round her,
the cat, understandably,
left home.

Dear Time

Forgive my impertinence,
but is it true that gravitation
is trying to worm its way
between you and space?

One hears rumors,
but, well, at close quarters
the strong force is a bully
and I don't trust him.

And maybe it's immaterial,
but while I'm being indiscreet,
please would you tell me
whether you are discrete?

It's one matter to learn
that you're relative,
quite another to think
you're particular.

For my final question,
if it's not too forward,
how is it that you're such
a straight arrow?

Always moving ahead,
doing what the universe says.
Never thrown for a loop.
No turning back. No regret.

The Headlong Year

If years had just eighty days,
Earth's orbit hastier than Mercury's,
would our tempers be equally precipitate,
our tweets a mere thirty characters?

—*This poem was first published in April 2017 when tweets were limited to 140 characters.*

The Year's Slow Turning

If years had ten thousand days,
Earth's orbit as stately as Saturn's,
would we chase after summer,
following the arrowing geese?

Or would we stay put,
tenacious as beavers,
hoarding our woodpiles,
our winter children?

The Red Planet

1. 1963

The Space Race won the hour
that Sputnik 23 transmitted
its first photo of the Martian surface,
showing in graphic clarity
a network of canals.

The Red Planet proclaimed
as the birthright
of every Soviet worker.
What worth the dead and barren Moon
beside a living world?

2. 1977

Alexei Leonov landed
on Hellas Planitia:
those temperate rolling plains,
areothermally warmed,
dense with rust-colored shrubs.

Leonov the first to touch the soil of Mars,
the first to walk its silent miles,
to plant the Red Banner
by a deserted canal
left by some long-vanished race.

Mars As It Might Be

I've been walking on Mars again,
walking the underground caverns
where he guards his gardens,

spaces tall as the palm trees
that stretch into leafy canopy,
new growth on an aging planet,

this green and secular cathedral
housing trees, orchids, spices,
hand-tended, hoarded, beloved.

I've been walking on Mars again,
or Mars as I imagine him
a hundred years from now.

A cold, almost airless planet,
the sun dimmed by distance,
his oceans desiccated.

Who came to him and why?
Who wrested ice from polar cap
to water a pride of palm trees?

I've been walking on Mars again,
that lightness to each footprint
due to his diminutive stature,

or due, perhaps, to hopes held
in common by those who came
and settled beneath his surface,

all the things they left behind
to ferry worms, the first soil,
nut, fruit, seed, sapling.

Holding You

In the dark,
I lifted you out of sleepiness
into my arms,
school ahead of you,
but for the minute
that I carried you
down the stairs
to breakfast,
you were a larger version
of your toddler-self,
your head on my shoulder,
your limbs still floppy;
the universe so vast
that if I walked
for seven years
I wouldn't even reach the Moon,
the Sun three thousand years beyond that
and the next nearest star
eight hundred million years away,
and in that vastness of length
only I had the privilege
of holding you.

ACKNOWLEDGMENTS

The earliest origins of these poems lie far back when my father taught me the names of the planets in order of their distance from the Sun. Along the way, a host of science fiction stories and a wealth of poetry laid down roots. And then there were my science teachers and uncounted non-fiction books and articles.

I am greatly indebted to all of these, but must single out one in particular—*The Daily Poet* by Kelli Russell Agodon and Martha Silano. The poetry prompt for September 14 on writing an unlikely "How To" poem launched me into the main sequence of this collection.

Many thanks to Lisa Rodgers, my agent, for her expertise, friendship, and general awesomeness.

Hip, hip, hooray to my family—Andrew for being the best of first readers, Lucy for her artwork and creative camaraderie, and William for sharing my love of space.

All honor to Diane Turnshek and F. J. Bergmann, proofreaders par excellence. Remaining infelicities, typos, and questionable comma choices are my fault, not theirs.

Abject gratitude to Jay O'Connell, longstanding friend, for his skill, patience, and for being cheeky on my behalf.

ORIGINAL PUBLICATION

Poems from this collection first appeared in the following publications:

"How to Paint Mercury," Apex Magazine
"How to Colonize Venus," Andromeda Spaceways
"How to Thank Earth," Uppagus
"How to Decorate the Moon," Eye to the Telescope
"How to Seduce Apophis," Abyss & Apex
"How to Tidy the Asteroids," Rune
"How to Question Asteroid 16 Psyche," Mithila Review
"How to Tease Jupiter," Uppagus
"How to Colonize Ganymede," NewMyths.com
"How to Surprise Saturn," Spillway
"How to Advertise Titan," Star*Line
"How to Fly By Neptune," Apex Magazine
"How to Speak to Pluto," Apex Magazine
"How to Grade Comets," Songs of Eretz Poetry Review
"How to Be a Star," Uppagus
"How to Stop Being a Star," Uppagus
"How to Mislay Constellations," Triangulation: Dark Skies
"How to Call Proxima Centauri," Star*Line
"How to Imagine Barnard's Star b," Star*Line
"How to Honor Sirius," Eternal Haunted Summer
"How to Weave the Stars," NewMyths.com
"How to Appoint the North Star," Frozen Wavelets
"How to Notice a Dark Nebula," Triangulation: Dark Skies
"How to Observe the Horsehead Nebula," Uppagus

"How to Mourn Kepler's Supernova," Penumbric
"How to Weigh Yourself," Rune
"How to Hide the Milky Way," Uppagus
"How to Betray Sagittarius A*," Strange Horizons
"How to Enlist in Andromeda," Dreams & Nightmares
"How to Glimpse the Icarus Star," Triangulation: Dark Skies
"How to Lie About SN 2213-1745," Mithila Review
"How to Construct the Elements," Asimov's Science Fiction
"How to Count Astronomically,"
    The Magazine of Fantasy & Science Fiction
"How to Trick a Trickster," Eye to the Telescope
"How to Fathom a Light-Year," Amazing Stories
"How to Commemorate the Big Bang," Uppagus
"How to Forfeit the Future," NewMyths.com
"How to Overlook Differences," Uppagus
"How to Curve Spacetime," Asimov's Science Fiction
"How to Emulate Black Holes," Star*Line
"How to Dance with Dark Matter," Uppagus
"How to Hear the Stars,"
    The Magazine of Fantasy & Science Fiction
"How to Detect Solar Neutrinos," Silver Blade
"How to Go Eighth," Triangulation: Dark Skies
"How to Go Twelfth," Analog Science Fiction & Fact
"How to Go Seventeenth," Triangulation: Dark Skies
"Old Astronauts," Star*Line
"The Third Man," Paper Street Online
"Félicette, October 18, 1963," Songs of Eretz Poetry Review
[Groundless, that first step], Twitter
"Dear Creator," The Magazine of Fantasy & Science Fiction
"The Cats of Mars," NewMyths.com
"The Headlong Year," Eye to the Telescope
"The Red Planet," Star*Line
"Holding You," Louisiana Literature

ABOUT THE AUTHOR

Mary Soon Lee was born and raised in London, but has lived in Pittsburgh for thirty years. She is a Grand Master of the Science Fiction & Fantasy Poetry Association, and once upon a time earned a degree in mathematics from Cambridge University. Her work has appeared in a wide range of publications including *American Scholar,* the *Magazine of Fantasy & Science Fiction,* the *Pittsburgh Post-Gazette,* and *Science.* She hides her online presence with a cryptically named website (marysoonlee.com) and an equally cryptic Twitter account (@MarySoonLee).

ALSO BY MARY SOON LEE

POETRY
The Sign of the Dragon
Elemental Haiku

STORY COLLECTIONS
Ebb Tides and Other Tales
Winter Shadows and Other Tales

Made in the USA
Middletown, DE
16 September 2023